DRAGONFLIES

DRAGONFLIES

Magnificent Creatures of Water, Air, and Land

PIETER VAN DOKKUM

Yale UNIVERSITY PRESS/NEW HAVEN & LONDON

Yale University Press books may be purchased in quantity for educational,
business, or promotional use. For information, please e-mail
sales.press@yale.edu (U.S. office) or sales@yaleup.co.uk (U.K. office).

Designed and set by Mary Valencia
in Adobe Garamond type.
Printed in China.

Library of Congress Cataloging-in-Publication Data
van Dokkum, Pieter, 1972
Dragonflies : magnificent creatures of water, air,
and land / Pieter van Dokkum.
        pages   cm
ISBN 978-0-300-19708-2 (alk. paper)
1. Dragonflies—United States. 2. Dragonflies—United States—
Pictorial works. I. Title.
QL520.2.A1D65 2015
595.7'33168—dc23   2014025844

A catalogue record for this book is available from the British Library.

This paper meets the requirements of ANSI/NISO Z39.48-1992
(Permanence of Paper).

10 9 8 7 6 5 4 3 2 1

To Nel Koen

CONTENTS

Common Green Darner

DRAGONFLIES

The Secret Pond

The idea of capturing the beauty and mystery of dragonflies in photographs had been with me for several years, but it was the discovery of the pond that somehow made it seem possible. The pond is small, oval, and rather unremarkable. It is bordered by trees and tucked in a corner of a small park in a small New England town. Aside from the pond the park features some hiking trails, a pen with farm animals, and a little playground, where my daughter learned how you can cheat gravity on a swing.

I do not know why so many dragonflies have decided to call this little pond home. At the height of summer hundreds zoom over the water, demonstrating their aerial prowess to rivals, members of the opposite sex, and whoever else happens to be looking. Over the years I learned to identify the species that live in the park and recognize their behavior, and I have come to know the little pond like the back of my hand. I have waded in it, lost shoes to its muddy bottom (to the consternation of families visiting the park), and spent nights along the water's edge waiting for nymphs to emerge (to the consternation of the park ranger).

After a while I began to see dragonflies everywhere—and so can you. They dart through our world, flying, seeing, hunting, mating, usually as oblivious of us as we are of them. They hover over parking lots, hunt in city parks, and visit our gardens. They are the true fairies in our lives: wondrous winged creatures that are seen in glimpses, from the corner of the eye. The aim of this book is to freeze these moments by means of photography, bringing dragonflies in for a close-up view.

Along with photographs, this book offers information on various aspects of dragonflies, with a focus on the biology of common pond and wetland species in the northeastern United States. About a third of the photographs in this book were taken at "my" little pond. The remainder come from various locations in Arizona, California, Connecticut, Florida, Maine, Massachusetts, Michigan, New Mexico, North Carolina, and The Netherlands.

The light of the full moon reflects off the water on this warm summer night. Amid a chorus of cicadas and frogs, an inconspicuous creature that has spent its entire life under water is dreaming of wings. The dragonfly nymph slowly clambers up a reed. Inside its hard shell everything has been prepared for this moment: wing buds have appeared on its back, a network of tubes for breathing air has been put in place, and a new set of compound eyes has developed behind the larval ones. After the nymph reaches the top of the reed, it is time for its great transformation, which will take the rest of the night. By morning, birds and other daytime predators will find nothing but a lifeless, empty shell still clinging to the reed as if to remind its former occupant of its dark and shadowy youth.

The life cycle of dragonflies is superficially similar to that of butterflies: they begin life as eggs; the eggs hatch into nymphs; and the nymphs transform into their adult form by way of a metamorphosis. Adults lay eggs, and the cycle starts again.

Whereas caterpillars are as cute as insect larvae can be, no one would use that word to describe dragonfly nymphs. They live in water, where they are among the most ferocious and (presumably) feared predators in the community of small fish, amphibians, and insects that live in ponds and streams. Dragonfly nymphs have large eyes and a huge, extendable lower lip that they use to capture prey: essentially anything that moves and is not too much bigger than they are. Unlike butterflies, dragonflies have no pupal stage: the adult emerges directly from the nymph in a span of hours. This amazing transformation is a good place to start this book.

Nymph of a Darner dragonfly. Dragonfly nymphs are the stuff of nightmares if you are a small aquatic creature. They have excellent vision and a unique extendable toothed lower lip (in some species equipped with claws), which they use to grab unsuspecting prey. Nymphs breathe through gills located inside the abdomen.

Nymph of a damselfly. The gills of damselfly nymphs are outside their bodies, at the tip of their abdomen, and resemble three finely veined fins.

Alien invasion? Nymphs, like adult dragonflies, share a basic body plan but show large variation between species. The nymph on the left is of the Skimmer family; the one on the right is of the Darner family.

Journey to the surface. When ready, the nymph looks for emergent vegetation and begins to climb, a journey that will take it irrevocably out of the familiar water and into the air. The nymph at left is a damselfly; the one on the right is a Darner dragonfly. Although its outward appearance has not changed very much, its metamorphosis is already quite far along: the nymph is now little more than a shell around an adult body. The wings are folded inside wing buds that appeared on the back in the last few moltings.

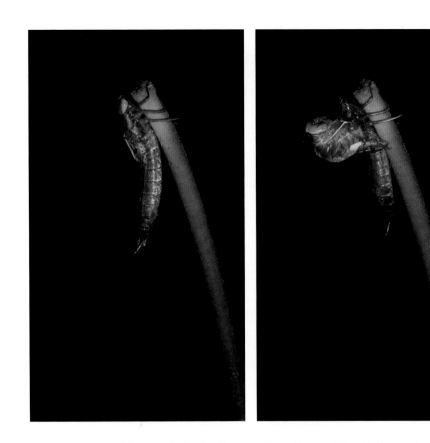

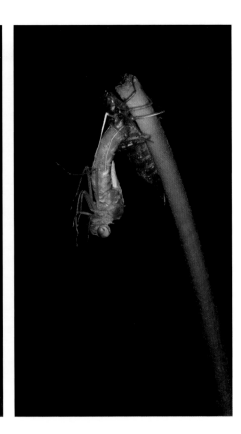

Metamorphosis of a Common Green Darner. After climbing to the top of a reed, the nymph bends its abdomen while breathing in air, cracking open the skin right below the head. The dragonfly emerges head first and upside down. Hanging from the nymph shell (exuviae), it has to wait until its legs harden before it is able to crawl out and turn around. In the space of about fifteen minutes, the dragonfly extends its wings by pumping air into them. Over the next three to four hours the abdomen slowly extends and hardens. Shortly before dawn it spreads its wings. When the first light of the sun reaches the pond, the newly emerged dragonfly takes off on its maiden voyage, leaving its old life (and skin) behind.

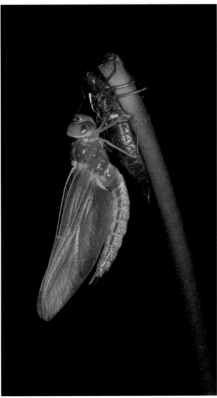

Birth of a dragonfly. In the midst of metamorphosis, this Common Green Darner's body has not yet hardened and is partially transparent. The light green pattern on the thorax will disappear in a few hours.

Metamorphosis of a Meadowhawk. The dragonfly has just emerged from his (now very small looking) larval skin. This is perhaps the most dangerous moment of the metamorphosis, as the creature is completely helpless.

The wings have extended but are not yet transparent, and the abdomen is still curled up. Recently emerged dragonflies are called tenera. Their coloring is usually pale and different from that of adults. Newly emerged males often resemble adult females.

Just-emerged dragonflies silhouetted against the full moon. Most dragonfly species emerge at night to avoid being eaten by birds and other daytime predators while their metamorphosis is in progress. On some nights hundreds of dragonflies may emerge from the same pond, a quiet and rarely witnessed spectacle. The wings are not yet transparent at this stage, and the interplay of moonlight and the intricate wing structure creates hues of blue and gold.

Facelift. Adult dragonflies show only a passing resemblance to their former selves.

Young damselfly. Most damselflies emerge at night, just like dragonflies, but their metamorphosis is typically faster. This damselfly flew off into the night about an hour after emerging from the water.

Early morning. The daunting metamorphosis is over, and after the morning dew evaporates the newly emerged dragonflies will use their wings for the first time.

First flight. Flying takes practice, and the first try is often a clumsy flutter to the ground. Many dragonflies fall victim to birds and spiders in these initial hours of life on the wing.

Empty shell. After the dragonfly has flown off, the exoskeleton of the nymph still clings to the reed. These exuviae can remain for days or even weeks.

An Excellent Design

I the course of the earth's history many different types of animals have come and gone, responding to changes in climate, geology, the availability of appropriate food sources, and the rise and fall of other species. However, some body plans are so successful that they have endured virtually unchanged for millions of years. The shark is a famous example, having similar-looking ancestors dating back 400 million years. The dragonfly is in the same illustrious group and is thought to be one of the first flying animals, appearing in the fossil record some 300 million years ago. The wingspan of some species reached an astonishing two and a half feet, probably owing to higher oxygen levels in the atmosphere and the absence of predators such as birds. These ancient giants were the largest insects that ever lived. If their nymphs were still around today we would probably think twice before going in the water!

Like all insects, dragonflies and damselflies have an external skeleton and six legs arranged in three pairs. Their bodies are composed of three parts: the head, which is dominated by a pair of huge eyes; the thorax, the power center to which the legs and two pairs of wings are attached; and the abdomen, which can bend, as it is made up of ten segments. The head is mostly used for seeing and eating, the thorax for movement, and the abdomen for breathing, processing food, and reproduction.

Dragonflies and damselflies are both members of the order Odonata, or "toothed ones," referring to their tooth-like mandibles. They have a similar body plan, but there are some important differences. Dragonflies are generally larger and stockier than damselflies. While resting, their wings remain spread, whereas many damselfly species hold them over their backs. Their eyes and head are also different: dragonflies have huge eyes that in many species touch at the top of their heads; damselflies have widely separated eyes. It is easy to overlook the humble damselfly as it slowly goes about its business inches above the water, while its more robust cousins are zooming overhead. Nevertheless, those who make the effort to observe these delicate, slender creatures will find they have a beauty—and a fierceness—all their own.

A proven design. This fossil dragonfly was found in Liaoning Province, China. It is from the Lower Cretaceous, about 130 million years ago. The dragonfly at right, a Black Saddlebags, is from present-day Connecticut. Dragonflies were among the first flying animals, and their body plan has changed very little in the past 300 million years. (Fossil image: The Virtual Fossil Museum, www.fossilmuseum.net)

Flame Skimmer (m)

Dragonfly (above) *and damselfly* (right). Damselflies are generally smaller and more slender than dragonflies. Dragonflies have broad wings, which they keep extended when perching. Damselflies have relatively small wings, which they hold backward when they are at rest.

Eastern Forktail (f)

Eyes of a dragonfly. The eyes of dragonflies are huge and often touch at the top of the head. They comprise some thirty thousand individual simple eyes (ommatidia) and provide their owners with better vision than any other insect. They are sensitive to a wide range of colors, including ultraviolet light. The small gray domes seen here in between the large compound eyes are ocelli: advanced light-sensitive organs that are probably important for stabilizing flight.

Eyes of a damselfly. The bulbous compound eyes of damselflies are widely separated and positioned on short stalks. Their field of vision is three hundred degrees, almost twice that of human eyes. The eyes often have a different color from the rest of the head.

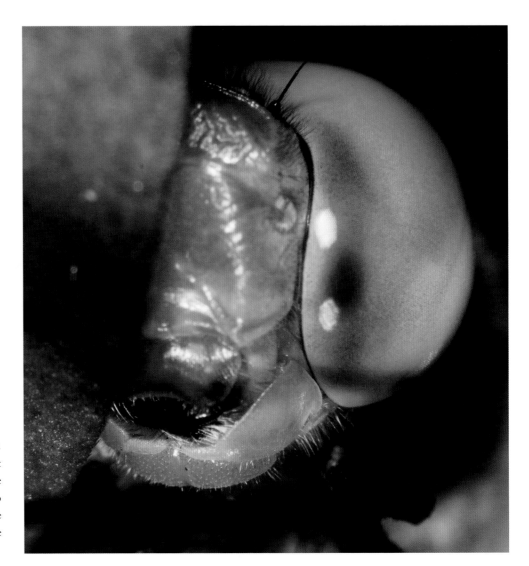

Eye and jaws of a young dragonfly. The eyes are still pale and have not yet acquired their final coloring. The dark spots (pseudopupils) seem to give direction to their vision; these spots absorb more light than the lighter parts of the eye.

Dew-covered eyes. The dew drops act as lenses, enlarging the six-sided facets at the tops of the ommatidia. Each ommatidium is an individual light detector, providing information that is akin to a single pixel of a digital image. Insect eyes are exquisite motion detectors, as adjacent ommatidia respond in sequence to even the smallest movement. About 80 percent of a dragonfly's brain power is devoted to processing visual information.

Stained glass wing. The wings have a complex, rigid surface that is maintained by a network of veins. The subtle colors of this immature Black Meadowhawk are caused by sunlight reflecting off the still not quite transparent wings.

Attachment of wings. Each of the four wings has a separate connection to the thorax, enabling dragonflies to move them independently from one another. Powerful muscles are attached to each wing base. These muscles can adjust the frequency and power of each stroke and also the angle of the wing.

Wing structure and color. Dragonfly wings are mostly transparent and generally do not have the beautiful colors of butterfly wings. Nevertheless, they often feature colorful spots and bands, which can be used to identify species. The veins in the wings can be colored, too. This photo shows a detail of the wing of a Calico Pennant, covered in dew.

Side view. Dragonfly wings are not flat but have a complex corrugated structure with ridges and valleys. This improves the stiffness of the wings and possibly also provides extra lift during flight.

Not for walking. Dragonflies have six long and powerful legs, equipped with barbs along their length and ending in a sharp hook. Their sole purpose is to grab things: reeds to rest on, prey, rivals, and mates. Some species, such as the Flame Skimmer above, fold their front legs when perching. Unusual for insects, dragonflies rarely walk.

Blue Dasher (m)

Segmented abdomen. The abdomen consists of segments that can move with respect to one another, enabling dragonflies to bend their bodies.

Thermoregulation. Dragonflies regulate their body temperature by angling their bodies to maximize or minimize the area exposed to the sun. When temperatures are high around midday, perching dragonflies such as this Halloween Pennant may point their abdomen straight up, to absorb as little heat as possible ("obelisking").

Breathing. The bodies of dragonflies contain a branching network of tubes, through which air is pumped. This pumping is achieved by expanding and contracting the abdomen, with a slightly faster rhythm than that of human breathing. In broad-bodied dragonflies such as the Common Whitetail (above), the width of the abdomen visibly increases and decreases (right) a few times per second. Watching a dragonfly breathe gives some sense of kinship with these otherworldly creatures.

Tip of the abdomen. The shape of the tip of the abdomen is often very different in males and females, as it is used for different purposes in the reproduction process. Among other things, males use the tip for clasping females, and females use it for laying eggs. This difference in shape is particularly clear in the appropriately named Clamp-tipped Emerald dragonfly; the male is on the left and the female is on the right.

THREE

Dew-clad Dragons

The photographs in this and the following chapters show dragonflies as they go about their daily activities. Because this is not a field guide, no attempt is made to cover all 5,500 dragonfly and damselfly species, or even the 450 species that occur in the United States. Instead, the photos aim to convey the intimate and subtle magic of dragonflies, and their brilliant colors, delicate wings, and spectacular flight. They offer a glimpse into a world of air and water and movement, occasionally interrupted by a short rest on a thin reed swaying in the breeze.

We begin on an early morning, just before sunrise. The pond is still and empty, its surface shrouded in mist. Dragonflies are nowhere to be seen, although they will soon take possession of the air. As the light began to fade the previous evening they sought out protected spots in trees and on plants, sometimes quite far from the water. There they still are, clinging to a stem or a leaf, awaiting the first rays of the morning sun.

Do dragonflies sleep? At night they are motionless, with all bodily functions greatly diminished. Because the physiology of insects is so different from that of mammals, scientists are reluctant to describe this state as sleep, and have settled on the word "torpor" instead—although an observer would be forgiven for thinking that a dew-covered dragonfly on a cold morning is fast asleep. When the sun comes up, the dragonfly will rub its eyes, removing the dew. It will slowly warm up, and as the dew evaporates and the wind picks up, it will take off for its first flight of the day.

Dew-covered dragonfly. On cold mornings in late summer insects can easily be approached and photographed because they are sluggish and often covered in dew.

Sparkling. The morning sun reflects off the dew drops on the wings of this female Saffron-winged Meadowhawk.

Ruby Meadowhawk (m)

Calico Pennant (f)

Front and back. Two views of the same Calico Pennant, just before it shook off the dew drops on its wings.

Before dawn. Spreadwing damselflies, just before sunrise.

Hide and seek. Damselflies are light sleepers—insofar as insects sleep at all. Even when they are cold and covered in dew, they tend to move away from whatever is approaching them and hide behind the plant they are clinging to. A curious eyeball may be the only part of the damselfly that is visible to predators—and cameras!

Watching the flight of dragonflies is mesmerizing. Although they are most easily observed when stationary, dragonflies really are creatures of the air, and their bodies are made for flying. They have a mastery of their element that is unique in the animal kingdom: they can fly upside down, stop, and change direction in the blink of an eye, hover, and suddenly accelerate to speeds of up to thirty miles per hour. Their flight appears erratic, but after a while patterns emerge: a short hover near the edge of their territory; a backward flip to grab a gnat out of the air; a sudden acceleration when a rival is spotted; a furious ball of legs, wings, and jaws when the rival comes too close. With more time, one can learn to follow their flight with a camera, stealing moments of an aerial display that has been honed to perfection over millions of years.

The flight of airplanes and birds is made possible by their curved wing shape, which generates upward pressure as air flows around it. Dragonflies use a different method: experiments have shown that they create and use turbulence in the air to get extra lift. The airflow around the wings is very complex, and its study may one day lead to new types of aircraft. Dragonflies are also able to move each of their wings independently of the other three, giving them great maneuverability. In normal flight the front wings usually lead the back wings, a method of flight called phased stroking. Other flight methods used by dragonflies are counter-stroking, with one pair of wings going down while the other goes up, and synchronized stroking, with the four wings going up and down together.

Most of the actively flying odonates above streams and ponds are male dragonflies, which are patrolling their territory or looking for females. Damselflies have relatively small wings and are much less accomplished fliers than dragonflies; they usually hover close to the surface and spend a lot of time perching on plants near the water's edge.

Blue-eyed Darner (m)

Dragonfly in flight. While flying, the dragonfly's head, thorax, and abdomen are all aligned. The legs are folded below the thorax.

Flight pattern. A common method of cruising flight is phased stroking, with the front wings leading the back wings by about half a beat.

Explosive takeoff. When perching dragonflies spot prey, they maximize their lift and acceleration by moving their four wings rapidly up and down, with little or no lag between the fore and hind wings. This series was taken with a high-speed camera; the time between successive frames is only 1/240 of a second, and the entire sequence is (literally) shorter than the blink of an eye. To our eyes dragonflies seem to disappear from their perch instantaneously. These explosive takeoffs help make dragonflies very successful hunters: most of the time they will be chewing on something when they return to their perch.

Common Green Darner. This large dragonfly is perhaps the most iconic of American species. They spend a lot of time on the wing, patrolling over ponds and hunting above meadows. Some populations of Common Green Darners are migratory, flying from the southern to the northern United States and Canada in the spring, with their offspring returning south in the autumn.

Twelve-spotted Skimmer. These are among the most striking and easily recognized dragonflies in North America. Males of this species often perch by the side of the water, taking off regularly for short flights to feed, fight, or find a mate.

Common Whitetail (f)

Black-tipped Darner (f)

Green-striped Darner (m)

Carolina Saddlebags

Carolina Saddlebags

Dragons or fairies? The transparent wings of dragonflies are usually difficult to see, but when lit up by the sun they can be a flash of fast-moving brilliance. Looking at dragonflies dancing in the light of the setting sun makes it easy to understand why they have often been associated with fairies.

Wandering Glider. These conspicuous yellow dragonflies are astonishing fliers: they can cross oceans and have been found all over the globe. It is the only odonate species on Easter Island, and it has been recorded above twenty thousand feet in the Himalayas.

Twelve-spotted Skimmers, male and female

Dragonflies are most easily appreciated while they perch on vegetation during the day. Many species of dragonflies and damselflies take short flights from a favorite stem or leaf that they call home for the day. While dragonflies that are perching may seem to be at rest, they are in fact constantly scanning the sky for potential prey, rivals, and mates. Many of the most easily found dragonflies that live near ponds or slow-moving waters are in the Skimmer family, which includes species with a wide range of sizes, body types, and colors. The large, powerful dragonflies in the Darner family are conspicuous and abundant as well, but they perch only occasionally, spending nearly all their time on the wing.

The photographs in this chapter showcase the variety of dragonflies as they may be encountered in the field. Although dragonflies need water to reproduce, they can be found in forests, mountains, and meadows quite far from a stream or pond. They sometimes perch on flowers; dragonflies have little use for them but are keenly interested in the insects they attract. Most of the species in these pages can be found and approached fairly easily, and many of the details in the photographs can be seen with the unaided eye. In most photos the species and sex are indicated. Males are generally more conspicuous in the field than females: in many common species they have brighter colors, and they are the ones actively patrolling their territories.

Cherry-faced Meadowhawk (f)

White-faced Meadowhawk (m)

Spangled Skimmer (m)

Widow Skimmer (m)

Eastern Amberwing (m)

Hidden variety. This dragonfly, a Unicorn Clubtail, looks somewhat different from many of the other dragonflies in these pages. It has a slender abdomen that gets thicker close to the tip (the "club" in Clubtail), and relatively small, wide-set eyes. Although there are many Clubtail species, they tend to be more elusive than the showy and common Skimmers and Darners.

Variegated Meadowhawk (m)

Twelve-spotted Skimmers (m)

Blue Dasher (f)

In hiding. Most dragonflies around the water's edge are males, who are busy patrolling their territories and looking for females. Meanwhile females tend to hide in nearby vegetation, only occasionally venturing out to the water.

Widow Skimmer (f)

Halloween Pennant (f)

Flame Skimmer (m)

Perching vertically. On warm days dragonflies in the Darner family fly more or less continuously. When they perch, they usually hang vertically from a branch or twig.

Common Green Darner (f)

Blue-eyed Darner (m)

Painted Skimmer (m)

Blue Dasher (m)

This female Common Green Darner fell in the water after an encounter with several males. She managed to "swim" to a piece of wood by letting her wings vibrate on the surface of the water. About fifteen minutes later she was airborne again.

Eastern Pondhawk (f)

Golden-winged Skimmer (m)

Roseate Skimmer (m)

Clamp-tipped Emerald (f)

Slaty Skimmer (m)

Black and yellow monster. The aptly named Dragonhunter is probably the most awe-inspiring American dragonfly. With a length of over three inches and long, powerful legs, it preys on butterflies and other large dragonflies.

Flame Skimmer (m)

Black Meadowhawk (f)

Female Seaside Dragonlet. Dragonflies typically require fresh water to reproduce. The Seaside Dragonlet is the only American dragonfly that breeds in salt water. It does not venture far from the coast, and may be found in salt marshes and tidal flats.

Ebony Jewelwing damselfly

Orange Bluet damselfly

Spotted Spreadwing damselfly

Emerald Spreadwing damselfly

Azure Bluet damselfly (m)

Orange Bluet (m)

Emerald Dragonfly, grabbing a mosquito.

Dragonflies and damselflies are carnivores and mainly eat other insects. They are not picky and will hunt any bug they can find. Mosquitoes, gnats, and other small insects form a substantial part of their diet, but they also catch larger prey such as butterflies and other damselflies and dragonflies. There is even a report in the scientific literature of a Common Green Darner dragonfly wrestling with a Ruby-throated Hummingbird! They are completely harmless to humans, however. They do not have any tool to sting with, and they have no desire to bite unless they are held. Even in that situation, most dragonflies would be incapable of piercing the skin.

Dragonflies grab their prey out of the air, with astonishing accuracy: it is estimated that 95 percent of attempts are successful, making them one of nature's most accomplished predators. The dragonfly attacks from below, tipping upward at the last moment and using its outstretched legs to grab the unsuspecting gnat or mosquito. Some species are nearly continuously on the wing, scanning their surroundings for movement. The large dragonflies lazily flying over soccer fields and picnic areas are usually members of the Darner family, and they are usually hunting. Other species perch on a branch or a leaf, ready to take off when they spot potential prey. They typically return to their perch after the hunt.

Small insects such as midges and mosquitoes are usually an in-flight snack. Once caught, they are held between the legs, which form a little basket, and chewed by the dragonfly's powerful jaws. Dragonflies eat often and fast; they can easily consume fifty or more mosquitoes in a day. Larger prey is carried to vegetation or the ground and consumed there. Although it is rare to see a dragonfly eating, the actual catch can fairly easily be observed: the dragonfly will seem to fly aimlessly, then suddenly change direction, flip, and continue on its flight. Most of the time, this means lunch has been served.

Ready to hunt. Having spotted movement with its large compound eyes, this Variegated Meadowhawk is about to take off. Flights typically last less than a minute, and the dragonfly will usually return to the same perch. Dragonflies that catch flying insects from a perch are called "salliers."

Blue Dasher eating small prey. Dragonflies mostly eat small flying insects, such as mosquitoes and gnats. They eat quickly; this tiny creature was consumed in seconds.

Pondhawks eating. Eastern Pondhawks (female, left, and immature male, right) are voracious predators, and do not hesitate to attack insects close to their own size. When they are on the ground they are often munching on something.

Damselfly eating. Because of their smaller wings, damselflies are poor fliers compared with dragonflies, and sometimes become their dinner. Nevertheless, they too are quite adept at catching insects, usually grabbing them from vegetation (a tactic known as "gleaning") rather than from the air.

Frosted Whiteface (m)

In-flight dinner service. A rare photograph of a dragonfly holding its prey in flight. The legs form a basket from which there is no escape for the hapless victim (probably a mosquito).

Dragonfly eating a damselfly. Large prey such as this cannot be eaten on the wing, so it is transported to a stem or the ground before being consumed.

Canada Darner, finishing a meal. Members of the Darner family do not hunt from a perch but fly almost continuously, a method called "hawking." They will land only when they catch prey that is too large to consume in-flight.

Variable Darner, eating a butterfly

SEVEN

The Next Generation

Just like many other animals, dragonflies spend a lot of time and energy to ensure the survival of their species. Males seem continuously preoccupied with finding a mate and compete aggressively, attacking rivals who enter their territories and keeping a constant eye out for females. Battles over females can be fierce, and even noisy: two males and a female can become a buzzing concentration of energy and motion as wings and legs beat against each other.

Before copulating, male dragonflies and damselflies grab the female behind her head, using claspers on the tip of the male's abdomen. The female then bends her abdomen, bringing the tip in contact with the abdomen of the male just behind the thorax. The couple now forms a wheel, which in damselflies is heart-shaped. Amazingly, many dragonfly species copulate on the wing, continuing their flight as a curious twelve-legged, eight-winged circular creature.

After copulating, females set out to lay eggs, typically hundreds at a time. Many different strategies are employed to enhance the chances that at least some of the offspring will survive. Female dragonflies in the Skimmer family usually deposit their eggs in the water and hope for the best, whereas Darners and damselflies typically carve slits in vegetation and place their eggs inside. Males have a different concern: they want to make sure they were the last to copulate with the female before she lays her eggs. Some males guard the female while she lays her eggs, hovering over her; others take no chances and hold on to the female behind her head. The eggs, floating in the water or neatly tucked away in a plant stem, are now on their own, their parents' focus once again on water, air, and movement.

Preparation. Male dragonflies have two sexual organs: on segment 2 (close to the thorax) and segment 10 (the tip of the abdomen). Before mating, the male transfers sperm from segment 10 to segment 2.

Ruby Meadowhawks

First embrace. The male has spotted a female and descends on her. He grabs her with his legs and flies off.

Wandering Gliders

Mating wheel. The male holds the female behind her head, using claspers on the tip of his abdomen. The female bends the tip of her abdomen up to retrieve sperm from the male's segment 2. All this is accomplished while flying!

Autumn Meadowhawks

Mating Meadowhawks. Mating does not only take place in the air. These Ruby Meadowhawks landed shortly after their first contact.

Blue Dashers

Dot-tailed Whitefaces

Unplanned landing. This pair of Clamp-tipped Emeralds fell to the ground while attempting to mate. The male helped the female up and they flew off together.

Mating damselflies. Damselfly mating is similar to that of dragonflies, but the way their bodies curve is different, often creating a mating wheel that is—appropriately—heart-shaped.

Emperor Dragonfly depositing eggs. Eggs are usually deposited right after mating. The strategies of the males and females vary between species. In Emperor Dragonflies, a large European species similar to the Common Green Darner of North America, the female deposits her eggs in submerged vegetation.

Finding a nest. Black-tipped Darners carefully insert their eggs into the stems of vegetation. The ovipositor, the egg-laying machinery at the tip of the female's abdomen, sports a sharp blade, which is used to cut open the plant stem.

Holding on. Common Green Darners remain in contact while the female lays eggs in vegetation below the water surface. These large dragonflies can often be seen flying in tandem, on the lookout for a good place to deposit eggs.

Water landing. Flying and laying eggs in tandem is not always easy, and the pair sometimes has trouble staying airborne. Amazingly, this male was able to lift the female out of the water again.

Catch and release. Black Saddlebags have developed a remarkably acrobatic method of egg-laying. Flying in tandem, the male and female occasionally separate so the female can deposit eggs in the water. The male snatches the female as soon as her abdomen comes out of the water, and the pair flies on. The yellowish mass beneath the end of the female's abdomen in the photograph at left is a clutch of eggs. In the photo at right the male has just let go of the female; both are reflected in the water.

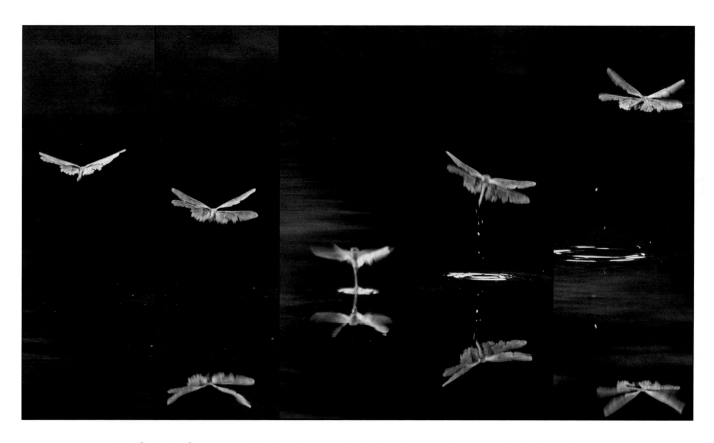

Egg-laying in the open water. The female repeatedly dips her abdomen in the water, each time depositing eggs. These photos capture a single dip, which takes about one second. The torn hind wings do not seem to bother this particular dragonfly too much!

Guarding an egg-laying female. Male Common Whitetails are devoted husbands and fathers—or as close to it as a dragonfly gets. The male hovers a few feet above the female while she is laying eggs, warding off attempts from other males to mate with her.

Aerial aggression. Two male Widow Skimmers are harassing a Common Whitetail pair.

A dragonfly nursery. Female Autumn Meadowhawks lay eggs in flight while being held by the male. Late in the season many pairs can sometimes be seen in a small area, depositing eggs by repeated tapping in partially submerged vegetation or mud.

Pearls of new life. This Ruby Meadowhawk is in the midst of laying eggs, allowing a rare peek at the tiny spheres that are being deposited in the water.

Reproduction in Eastern Amberwings. These small dragonflies have territories that often span only a few yards near the water's edge, which means their mating behavior can readily be observed. In this photo, a male looks on as another male is the first to find and grab a female.

1. The male pins down the female and grabs her with the claspers on the tip of his abdomen.

2. Once the female is successfully grasped, the male flies off with her to find a place to mate.

3. The pair forms the signature mating-wheel.

4. After mating, the male and female split up. The female goes back to her favorite branch and lays eggs, dipping her abdomen in the water.

Eastern Amberwing, just
after depositing eggs

Northern Spreadwings (f)
depositing eggs

Damselfly nursery. Some plants are very popular
with egg-laying damselflies.

Ovipositor. The egg-laying apparatus of a Spreadwing damselfly. The dark red blade on top cuts into the stem, after which an egg is inserted.

Damselfly eggs. The eggs are elongated and only about 1 millimeter long. They will hatch into tiny nymphs in a few weeks.

Dragonflies and damselflies do not live long: the odonate life cycle in temperate climates is usually a year. Most of this time is spent underwater as nymphs, whose bodies grow and change in successive moltings. Odonates do not shed their skin in their adult stage, making do with the bodies they obtained in their metamorphosis. These bodies, and particularly the delicate wings, are not made to last. The wings get damaged and torn in fights, and it is not unusual to see a dragonfly whose wings are shredded so much that flight no longer seems possible. The lifespan of an adult dragonfly is typically only a few months. In northern climates the lives of dragonflies can also be cut short by the onset of cold weather in the fall: odonates need warmth to power their muscles.

Old age is, of course, not the only cause of death. Dragonfly nymphs are an important food source for fish, crayfish, and water birds. Many dragonflies perish during metamorphosis, falling victim to spiders and other predators during this time of complete helplessness. Birds are often attracted to ponds when dragonflies are emerging in large numbers, and in the early morning hours they feast on the immature fluttering insects. Although adult dragonflies are superb fliers, they too fall prey to birds, such as Barn Swallows.

Individual dragonflies are ethereal creatures, whose wings beat only a short time in this world. Look closely, however, and just a few feet from the floating body of the dead dragonfly a damselfly is laying eggs. The life cycle turns rapidly for odonates, in the air and unseen beneath the surface of the water. Even in the dead of winter a nymph is scurrying on the bottom of a frozen pond, preparing for a future of light, air, and warm summer wind flowing over gossamer wings.

Problems during metamorphosis. A lot of things have to go right for metamorphosis to be successful. Many developmental processes have to be carefully choreographed and flawlessly executed. Sometimes dragonflies simply get stuck, dangling helplessly as their former skin becomes their prison.

Cedar Waxwings catching just-emerged dragonflies. Several of these birds showed up after a mass emergence of Meadowhawks. The Waxwings caught a large number of the insects as they fluttered near the ground, trying out their new wings.

Spider eating a Four-spotted Skimmer. It is difficult to feel sympathetic toward spiders if one is fond of dragonflies. Spiders may attack emerging dragonflies, and adults can get caught in webs. This spider is busily wrapping up its enormous catch.

Leftovers. Remains of a dragonfly, which was probably eaten by a bird. The predator did not bother with the wings or the abdomen; about 70 percent of a dragonfly's body mass is in the thorax.

Crashed. This immature male Eastern Pondhawk did not die of old age; it may have mistaken the duck-weed-covered water surface for the ground.

Spangled Skimmer

Blue Dasher

Structural damage. Despite their strength and agility, dragon-flies are delicate creatures and lack mechanisms to repair damage to their bodies. Their wings are particularly vulnerable to tears and cuts, and late in the season many dragonflies have a distinctly ragged appearance.

Faded brilliance. In death the colors of dragonflies quickly fade to duller tones. The eyes of the living Eastern Pondhawk here are colorful and seem very aware of their surroundings; those of the unfortunate one at right have turned a dark brown.

The end. In New England the colors of fall mark the end of the dragonfly season. Although they can withstand an occasional cold night, odonates need high temperatures to be active. A long cold spell will cause them to freeze or, more likely, starve. This is probably what happened to the female Meadowhawk at right, which ended up on the forest floor.

Life continues. Beneath the water surface odonates are hatching. This dragonfly nymph is only a few millimeters long and is still transparent; it has probably molted only a few times since hatching. There are no wing buds yet, and there is little indication of the life that awaits this little creature. However, unknowingly, but irrevocably, it has already begun to prepare for a far-off night filled with moonlight and the sounds of summer.

The sun on the hill forgot to die,
And the lilies revived, and the dragon-fly
Came back to dream on the river.

[from "A Musical Instrument,"
by Elizabeth Barrett Browning]

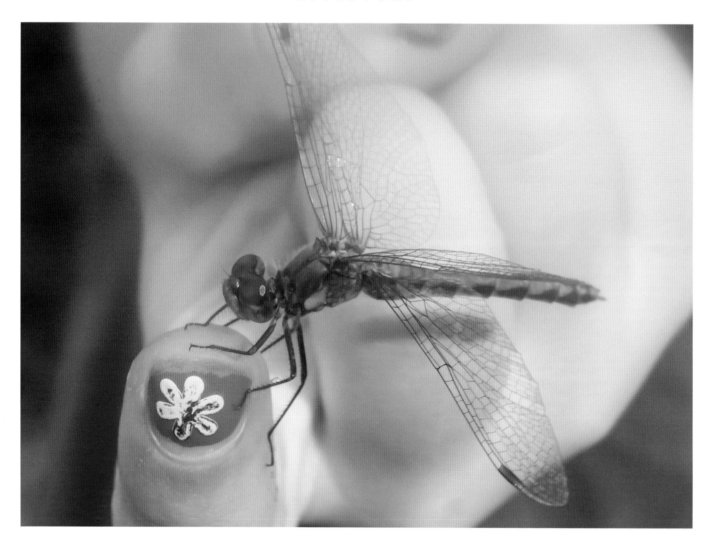

Dragonfly photography is available to anyone. Common species such as the Twelve-spotted Skimmer and Blue Dasher are quite conspicuous and tend to perch in easily accessible vegetation, bordering ponds and streams. If care is taken to avoid sudden movements, it is generally possible to approach within a few feet of them, which is close enough for frame-filling images. If they fly off it pays to wait a few minutes: many dragonflies tend to return to the same perch after a short flight. With patience the behavior of different species can be recognized and anticipated, and the quality of photographs will steadily improve.

Although any camera or mobile phone can be used to take recognizable dragonfly pictures, obtaining high-quality photographs is much easier with a DSLR. Almost all the photos in this book were taken with a Canon 40D or 5D Mark III camera combined with a 100 mm f/2.8 macro lens or a 300 mm f/4L telephoto lens. The main illumination was usually the natural light, with a dual-head macro flash unit sometimes providing fill light. A tripod was used only for the underwater, night, and early morning photographs; in general dragonfly photography is best done hand-held so one can respond to rapidly changing circumstances and compositions. The most challenging photos in this book are those of dragonflies in flight. They require a very fast shutter speed to freeze the action and a small aperture to obtain sufficient depth of field. Furthermore, all flight photos in this book were focused manually as even very sophisticated autofocus systems cannot cope well with these fast-moving, small subjects.

We, and our children, can continue to enjoy dragonflies only if they continue to share our environment. They need clean fresh water to thrive, and the reality is that many North American species are in decline due to habitat loss and water quality degradation. Dragonflies rely on our stewardship of marshes, rivers, swamps, and ponds—including the one described in the opening chapter of this book. The abundance of dragonflies living in and near "my" pond demonstrates that even urban areas can support healthy aquatic habitats when properly managed. It is my hope that this book encourages you to find your own secret pond, and experience the magic that these pages have attempted to capture.

There are quite a few odonate field guides, which are indispensable for identifying species. A good starting point is Blair Nikula, Donald Stokes, Lillian Stokes, and Jackie Sones, *Stokes Beginner's Guide to Dragonflies* (Little, Brown, and Company, 2002). Examples of guides focusing on specific regions are Steve Gordon, *Dragonflies and Damselflies of Oregon: A Field Guide* (Oregon State University Press, 2011), and Timothy Manolis, *Dragonflies and Damselflies of California* (University of California Press, 2003). A field guide covering all of North America is Sidney W. Dunkle, *Dragonflies Through Binoculars: A Field Guide to Dragonflies of North America* (Oxford University Press, 2000).

The most comprehensive illustrated field guides for North America, and excellent sources of information on dragonfly biology and behavior, are Dennis Paulson, *Dragonflies and Damselflies of the West* (Princeton University Press, 2009), and its companion, Dennis Paulson, *Dragonflies and Damselflies of the East* (Princeton University Press, 2011).

For those interested in odonate biology, Philip S. Corbet, *Dragonflies: Behavior and Ecology of Odonata* (Cornell University Press, 1999), is a standard text. Detailed information on North American species can be found in James G. Needham, Minter J. Westfall, Jr., and Michael L. May, *Dragonflies of North America,* revised edition (Scientific Publishers, 2000), and Minter J. Westfall, Jr., Michael L. May, and Sidney W. Dunkle, *Damselflies of North America,* revised edition (Scientific Publishers, 2006). Jill Silsby, *Dragonflies of the World* (Smithsonian, 2001), provides an in-depth discussion of dragonfly anatomy and physiology, as well as a worldwide overview of all odonate families and many species.

A more hands-on approach is offered by Forrest Mitchell and James Lasswell, *A Dazzle of Dragonflies* (Texas A&M University Press, 2005). This book covers various ways to turn interest in dragonflies into a hobby.

INDEX

Page numbers in italics refer to photographs